Space
at the
Speed
of Light

Space at the Speed of Lig

Space
at the
Speed
of Light

DR. BECKY SMETHURST

The History of 14 Billion Years for People Short on Time

TEN SPEED PRESS
California | New York

To you, whoever you are, for being curious
enough not only to pick up this book, but
also to open it.

Oh, and to Dad, for reminding me not
to become an accountant.

Contents

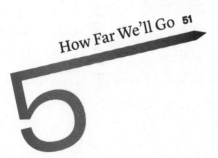

Preface

The wondrous thing about science is that nobody knows
the right answer. This is not how we are taught science as
children, though. In the classroom, theories are presented
as hard facts that have always been understood in that way.
Fortunately, the reality is far more creative: being a scien-
tist is like fitting in the pieces of an ever-changing jigsaw
puzzle for which we've lost the lid. The work of so many
people over decades, even centuries, slowly builds a picture
of our current best understanding. Although some areas of
science still have the odd piece missing, some others have
giant gaps for which we currently don't have the tools,

mathematics, or data even to glimpse what shape the pieces are.

Science is all about posing the questions to which nobody yet knows the answers. Convincing people there is a "right" answer—based on evidence and facts you, your colleagues, and your predecessors have collected to build a theory of the previously unexplained—is the crucial bit. This means that science moves quickly, with theories maturing and sometimes even boomeranging as more evidence comes to light.

The theories and facts outlined in this book offer a history of ten big questions about space and how science currently answers them. They are all considered successes now, but who knows how they shall be perceived in fifty years. Perhaps our current theory of dark matter will be scoffed at by future generations, akin to our current disbelief that learned minds once thought the Earth was at the center of the universe or that the atom could not be split. Yet, that doesn't mean we shouldn't treasure this knowledge now and the wonders of our world that it lays bare.

The chapters in this book cover the essentials behind the evolution of some of the most successful theories that describe the weird and wonderful objects in space, either for those who want new glimpses into its depths or have no prior knowledge of the secrets it contains. Reading this book will take you on a tour of the universe—from its beginnings in the big bang to the elusiveness of dark matter to a thoughtful consideration of whether life exists beyond our planet. If we linger when we reach black holes, that's because they are where my heart truly lies. My own scientific jigsaw puzzle, which I attempt while sat at my desk in the astrophysics department in Oxford, is helping to understand how these enigmas affect the galaxies they reside in.

We end with what we still don't know: the biggest question of all and one we will never be certain we've answered wholly or correctly. But, as an astronomer, this is the most exciting quest of all—pushing the boundaries of our knowledge, bit by bit, to uncover a fuller picture of the universe and our place within it. My hope is that this book will give you glimpses of that, as yet incomplete, masterpiece.

1 Why Gravity Matters

The Sun is just one star in more than 100 billion stars in our galaxy, the Milky Way. It's an island of gas, dust, and stars, more than a million trillion kilometers across. At the center of the Milky Way star system there's a black hole four million times more massive than the Sun. That's something we call a supermassive black hole, and like the Sun in our solar system, it's in the gravitational driving seat of our whole galaxy.

Isaac Newton discovered the law that governs gravity some centuries ago: two objects will attract each other proportional to how heavy each object is. The lighter object will be affected more by the force between the two objects. The force also depends on how far apart the two objects are: the farther apart they are, the weaker the force between them. If you double their distance, the force drops to a quarter of its original strength. With these laws, we can work out the effect of gravity between any two objects in the universe, including between you and the Earth under your feet.[1]

---

1 In case you're wondering, you feel a constant force of about 500 to 1,000 Newtons (depending on your weight) from the Earth's gravity pulling on your body. For context, the average force of a human bite is about 700 Newtons; a great white shark bite has a force of 18,000 Newtons!

The law of gravity brings order from chaos; it is, after all, what produced our solar system. Before the Sun formed there was only a giant cloud of hydrogen and helium gas, with a sprinkling of heavier elements like oxygen, carbon, and iron, left over from a previous generation of stars. This cloud was a swirling mess containing the atoms of each element. As each atom is a tiny particle of a certain mass, they were gravitationally attracted to all the other particles in that messy gas cloud. Those particles began to clump together under gravity, with the biggest clumps attracting the other clumps, until eventually gravity overcame the energy of all the particles whizzing around and trapped them together to, effectively, cool them down. The next step was the collapse of the gas cloud down to a really high density where the pressure increased so much that it got hot enough to ignite nuclear fusion and our star was born.

Nuclear fusion is when stars, like the Sun, turn four atoms of hydrogen into one atom of helium, and it's why all the stars in our night sky shine. So, what once was a swirling cloud of gas, full of atoms whizzing around, became a burning protostar because of gravity.

Now, that endlessly whirling cloud of gas will also have been harboring a remnant of the past. It will have inherited some leftover rotational energy, which we call angular momentum, from a previous generation of stars, perhaps even the first stars to form after the creation of the universe. This means that, overall, the gas cloud was swirling preferentially in a certain direction, so that as particles started to clump together under gravity, they assumed that same preference: the proto-Sun began spinning. What happened to the rest of the gas cloud around the early Sun is the same as what happens to a ball of pizza dough when you spin it above your head: it flattened out into a saucer, or a disk, which went on spinning. Inside the disk, the gravitational attraction between particles continued, so that more clumps formed into protoplanets around the Sun, giving us a beautifully ordered system where the planets (plus comets, asteroids, and other leftover bits of rock) all orbit around it in the same direction. This process is how we think all stars, not only the Sun, have formed.

The same thing is reflected in our own Earth-Moon system. The Earth spins in the same direction as it orbits because the tiny particles that clumped together to make it had inherited that hint of angular momentum from the previous generation of stars. Similarly, the Moon orbits around the Earth in the direction that the Earth spins.

But that's where the similarities stop, because the rest of the Moon's properties are consistently strange. Its day is as long as its year. That means the time it takes the Moon to spin around on its axis, its day, is the same time it takes for it to orbit the Earth, its year, which is twenty-eight Earth days. If the Earth followed this example, half the planet would always be in daylight and half would always be in darkness throughout the whole year, as the Earth made one orbit around the Sun. The Earth would spin at the same rate to keep one side always facing away from the Sun. This is why we only ever see one side of the Moon—we never see its far side because it never points toward us. That's not to say the far side is the dark side, however, because the Earth doesn't light up the Moon, the Sun does. This is why we see the phases of the Moon: We see a Full Moon when it is on the opposite side of the sky to the Sun, which then fully lights up the side of the Moon pointing toward us. And we see a New Moon when the Moon is between us and the Sun, so the Sun is lighting up the side that points away from us.

If you are now wondering why we don't get a total solar eclipse every twenty-eight days, considering the Moon passes between the Sun and the Earth on each of its orbits, the reason is because the Moon is not orbiting in the same plane that the Earth orbits around the Sun. It's slightly tilted by about five degrees. So, sometimes it passes just below, and sometimes just above, the Sun in the sky during its New Moon phase.

All of these properties of our Earth-Moon system might seem like happenstance, but they help reveal something about how our Moon formed. You might expect that the Moon formed around the Earth in similar

Phases of the Moon depend on the relative positions of the Earth, Moon, and Sun.

Seasons on Earth are a result of its tilted axis.

circumstances to how the Earth formed around the Sun, that is, from the leftover bits that didn't go into forming the Earth. But, our best theory for how the Moon formed is far more dramatic than that. It's called the giant impact hypothesis and it says that another protoplanet orbiting the Sun collided with the proto-Earth in the early days of the solar system. This impact liquefied the colliding planet and about half of the Earth because the energy in the impact was so huge. All of this liquefied planet rock was flung out into space while the Earth recovered and kept spinning. But, again, that liquefied rock couldn't escape the pull of gravity from Earth and thus it was pulled into a spinning disk that later clumped together to form the Moon.

This theory explains why the axis that the Earth spins around is off-kilter. In that collision, the Earth got a knock sideways so that it spins slightly on its side, a tilt of about 23 degrees, like a dog lovably cocking its head at you. This means that throughout the year, as the Earth orbits the Sun, during the Southern Hemisphere summer, the South Pole faces toward the Sun, and six months later the North Pole points toward the Sun. This is what causes our seasons on Earth. It's hotter when our hemisphere happens to be pointing toward the Sun due to the Earth's tilt.

It's incredible how much order and calmness you can produce out of so much chaos with one simple law of physics. The same law that causes apples to fall out of trees, keeps our feet on the ground, and causes our changing seasons affects everything in the galaxy and the solar system. And, it's not only in our own stellar backyard that we see this. Outside the Milky Way, we see even more of these islands of stars, of all shapes and sizes, in every direction we look in the universe. Gravity has formed them all from a huge cloud of chaotic hydrogen gas particles into ordered systems with beautiful spiral structures.

Although gravity has made these beautiful islands of stars, it can destroy them, too. Most galaxies aren't found in isolation, they're clustered together in groups bound by gravity. Our Milky Way, along with Andromeda, is part of the Local Group of galaxies. These are the two biggest galaxies in the group and so they are both gravitationally attracted to each other. One day, in the next four billion years or so, the Milky Way and Andromeda will collide with each other and the gravitational forces between the two will tear them both apart, disrupting the orbits of all the stars, until everything has settled back down into a giant blob of a remnant galaxy, Milkomeda.

This is an example of another law of physics, the second law of thermodynamics, which states that the entropy of a system can never decrease over time. *Entropy* is a measure of how disordered a system is; it looks at how randomly the particles in a system are moving. So, the universe as a whole is destined to become more random with the inexorable march of time. In four or five billion years, our Sun will run out of fuel and engulf the solar system, reducing all of the stuff in it to one more chaotic cloud of gas. The stars in our Milky Way galaxy will ultimately find themselves on chaotic, random orbits around the center of the galactic blob that will be Milkomeda. Such is the fate of all matter in the entire universe. Even as the laws of physics create order, that order will inevitably descend back into chaos once more.

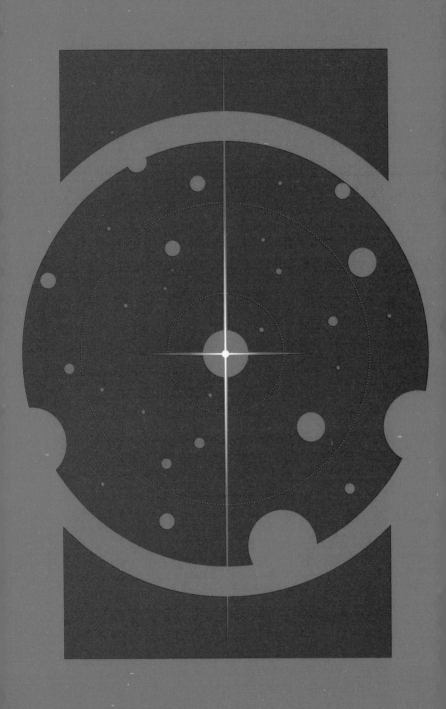

2

In
the
Beginning
There
Was

Nothing

The human brain cannot truly comprehend what "nothing" is. Simply by thinking of nothing, we turn it into something. So, when we say that there was truly nothing "before" the big bang,[1] it's hard for the brain to understand what that means. The concept of a "before" also doesn't exist without the big bang, as time itself was created in it. Even space didn't exist "before" the big bang, so there was "nowhere" that "nothing" could be found. All the energy and matter in the universe were produced in the big bang, and as a single human being, we are just one infinitesimal chunk of that energy budget. Energy cannot be created or destroyed according to the first law of thermodynamics, so we really are only working with what the big bang gave us—there is no mythical energy tree in the galactic back garden we can go out and shake when we have need of more.

The story of how we know the big bang happened is, perhaps, one of my favorites in the history of astronomy.

---

[1] Ironically, the big bang was neither big nor a bang. The name was coined back in the 1930s as a slur against the newly emerging hypothesis of an expanding universe, and it's been plaguing astronomers ever since.

Overnight, the edge of the entire known universe went from the farthest star at about a hundred thousand light years away, to containing whole other islands of billions of stars, trillions of light years away. The breakthrough came from a chap called Edwin Hubble in 1929[2] (building on work done by Shapley, Curtis, and Öpik earlier in the decade) when his observations revealed the existence of other galaxies beyond our own. Never has there been another result with a bigger impact on astronomy. In one fell swoop, Hubble paved the way for an uncountable number of fields of research and changed our view of our place in the universe.

I would have loved to have lived through that time—a time when, as one, humanity's entire worldview changed. Akin to the European explorers standing at the helm of their ships as the shores of America came into view five hundred years ago, when their known world suddenly catapulted forward—but on the biggest scale you can imagine. A scale that, even now, our brains struggle to comprehend.

Discovering that the universe was dramatically larger and older than anyone imagined was probably not something Hubble ever expected to do after focusing mostly on sports in his early life and then training as a lawyer. But, after turning his attention to his passion for astronomy,

2   Lemaître had theoretically predicted an expanding universe two years earlier in a French journal, but it fell into obscurity for most of the twentieth century. There's now a movement to get Lemaître's work more widely recognized.

he took a position at an observatory in California, and it was there that he observed what at the time were referred to as "spiral nebulae." A nebula, back then, was basically anything in the sky that clearly wasn't a star. They were the fuzzy, dusty things that umbrella-ed everything from star-forming regions in our own galaxy to the remnants of stars that had gone supernova[3] and, unknown at the time, galaxies of billions of stars. Hubble was trying to work out the distance to these "spiral nebulae" using stars that pulsated, changing their brightness.[4] These stars are called Cepheid Variables and were intensely studied in the Milky Way by Henrietta Leavitt in the early twentieth century.[5] She discovered that how often the stars pulsed was related to how bright they are, something we now call the Leavitt Law. The faster a star pulses, the brighter it shines.

---

[3] This is when a star runs out of fuel (hydrogen gas) to burn and starts to collapse inward under gravity. The outer layers of gas rebound off the star's core in a huge outward explosion.

[4] Gravity is constantly compressing stars and pulling material inward, whereas the energy released by the nuclear reactions taking place in the center of the star are constantly pushing outward. The interaction of these two processes can cause some stars to pulse.

[5] Sadly, Leavitt didn't live to see Hubble use her work to prove the universe was expanding, or gain much recognition for it while she was alive. Today, all distances measured in the universe are, essentially, calibrated against distances measured from Cepheid Variable stars. The scale of the universe literally rests on Leavitt's shoulders.

When Hubble spotted these pulsating stars in the "spiral nebula," he needed to understand how fast they pulsed in order to know how bright they were—and then from how bright they appeared he would know how distant the star was. This is the same thing that you do when you cross a road at night; based on how bright a car's headlights appear, you can tell how far away it is. From this, Hubble found that the Cepheid Variables in the "spiral nebulae" were more distant than anyone had ever considered, meaning the sizes of these spiral nebulae were as big as the Milky Way itself. What Hubble had actually observed were galaxies.

Hubble didn't stop there though. He also took what's called a spectrum of each galaxy. This is where you split light through a prism and look for the traces left by different elements in the rainbow of colors produced. If you're picturing Pink Floyd's *The Dark Side of the Moon* album cover, then you're on the right track. These traces left by different elements are a quantum physics fingerprint of sorts, always at the same color (i.e., wavelength) of light no matter what. What Hubble noticed was that these signatures all had their wavelengths shifted to redder colors, which meant they were moving away from Earth. Not only that, the more distant the galaxy was from Earth, the faster it was moving away from us. It was only natural to think that if everything is moving away from us, then, at some point, everything was much closer. Keep rewinding time and

Light's spectrum shifts to redder colors in galaxies that are moving away from Earth.

you'll eventually end up at a point where all the matter and energy in the universe is condensed into a single point. Thus the idea of the big bang was born.

This concept of an expanding universe is confusing to most people. For one, it gives us earthlings a distorted sense of self-importance. All the things in the universe are moving away from us. We must therefore be in a place of great importance, surely? Well, no. It's all a matter of perspective. Imagine the playground game Cat's Cradle, where you wrap an elastic band around each hand and then pull them apart to perform all manner of dexterous rope-based tricks to wow your friends. If you were to look down at your hands, you would see that as you pull your hands apart both ends of the elastic move outward. Now picture how it might look to a tiny observer sitting on your left hand. They can see the right hand moving away from them, but they can't feel themselves moving, any more than we can feel the Earth beneath our feet moving around the Sun or spinning on its axis. Likewise, a tiny observer on the right hand would only see the left hand moving away from them.

This analogy works equally well when you consider you haven't created anything new in the process. You have the same amount of elastic band you started with—it's just that the band itself has stretched. Similarly, the galaxies aren't the ones moving, space itself is expanding, but— here's the crucial bit—we're not creating more space as we

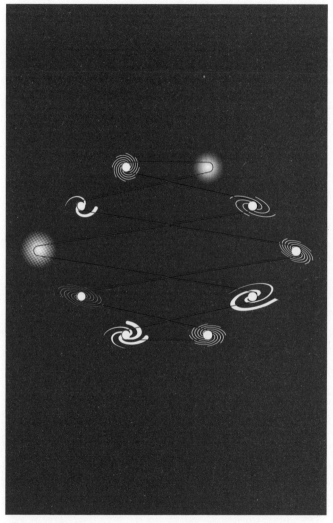

The expansion of space can be thought of like galaxies caught in a game of Cat's Cradle.

Infinite Expansion

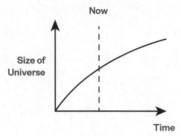

Stable Equilibrium

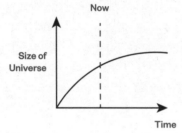

The Big Crunch

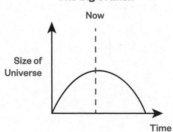

Sometimes scientists convey big questions using small graphs. These show three possible futures for an expanding universe.

do it. We are merely working with the space that the big bang gave us 13.8 billion years ago.

But let's not dwell in the past; it's far more exciting to think what will happen in the future—although, unfortunately, there will be no Hollywood ending to the universe. The first option is that we keep on accelerating in our expansion until there are such great distances between galaxies and stars that we are isolated in the vastness of the universe. The second, the Goldilocks universe, is more palatable. In this version, the expansion slows down until the universe reaches a happy medium: an equilibrium between the forces of gravity endlessly pulling inward and the inexorable pull of expansion outward. The third option is at the Shakespearian tragedy end of the spectrum: gravity ultimately wins the war against the expansion of space and the universe starts contracting inward until it reaches that tiny point where all the energy there ever was is reeled back into a "big crunch." Perhaps the universe has been doing this for eternity, cycling through big bangs and big crunches, with the odd waste product of carbon made in supernova occasionally producing life intelligent enough to contemplate this endless cycle.

A Brief History of Black Holes

3

A black hole is an object so dense that light can't escape from it. This means that the velocity you would need to escape the gravitational pull of that object would be greater than the speed of light. To understand this properly we have to understand gravity a bit differently from how Newton thought about it. This is what Einstein's theory of general relativity manages to do so neatly. Einstein said that wherever there is an object in the universe, it bends space-time[1] around it. You can think of this in two dimensions, like putting a soccer ball in the middle of a trampoline. The soccer ball in the middle weighs down the trampoline and curves it so it's no longer flat. If you then try to roll something like a ping-pong ball across the surface of the trampoline, it won't go in a straight line; its path will get curved as it travels because of the presence of the soccer ball. This is exactly how Einstein explained gravity, except that the

---

[1] Spacetime is a well-used word in a physicist's lexicon. It ties together the three dimensions of space (left and right, forward and backward, up and down) and one dimension of time. Einstein tied space and time together because he also realized that the faster you move through space, the more time slows down. That's his theory of special relativity, rather than general relativity.

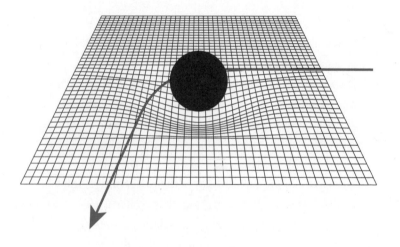

Some objects are so massive they warp the space around them. The red path shows how a less-massive object's trajectory could change near these warped areas of space.

soccer balls of the universe are the stars, the trampoline is space itself, and the ping-pong balls are the planets stuck going in circles around their stars because of curved space.

Now imagine a black hole bending space so much that, at least in two dimensions, you'd get to a point where space has been bent completely vertical: the black hole's *event horizon*. Any object—be it a spacecraft, ping-pong ball, or tiny photon of light—would never be able to recover after crossing the point of no return on the trampoline of the universe.

The official word for this maximum bending of space in general relativity is a *singularity*, where you have a huge mass in an infinitely small space. The idea of these infinite singularities that bend space so much infuriated scientists throughout most of the early twentieth century because it was messy mathematics. It means you have to divide by 0, which means the limit does not exist.[2] Eventually, Stephen Hawking and Roger Penrose came along in the 1960s and said that singularities were actually expected, given the known laws of physics, despite our issues understanding the math.

At this point, though, black holes were still considered theoretical objects—that is until Jocelyn Bell Burnell discovered neutron stars in the late 1960s (in the form of rapidly spinning neutron stars called pulsars that give out

---

2   This is what Lindsay Lohan's character in the 2004 film *Mean Girls* correctly identifies as the answer to the question at the end of the mathletes competition.

radio waves). Before this, astronomers knew that there was a physical mass limit beyond which a white dwarf star (held up by the repelling forces between electrons) would have to collapse into an entirely theoretical neutron star (held up by the repelling forces between neutrons). Until the discovery of neutron stars, no one had considered that these theoretical objects could actually exist in the universe. And if neutron stars were real objects, then what would happen when they too grew so massive that the repelling forces between neutrons also collapsed under the crushing force of gravity? The only logical conclusion was that they would collapse into a black hole.

Black holes definitely exist then. But what are they made of? The short answer is that we still don't really know. A neutron star is neutrons as densely packed as they can be in a crystal-like structure; so, what happens when the forces keeping them from being squashed are overcome? The simplest answer is: matter. In its most dense form, even if we don't quite know what that is yet.

But how do we know black holes are there if once they've formed we can no longer detect them with light? In the same way we can't see the wind but we see its effect on other things, like trees, sometimes we can see the gravitational effect of black holes on other objects in the universe. We've seen the light of background galaxies bend around interloping stellar-mass black holes in our own galaxy. This

Distant light bends around a stellar-mass black hole.

bends the light of the galaxy into a different shape due to the way light travels across Einstein's curved spacetime. We can't "see" the object that's curving the space but we can still tell it's there.

Another way to "see" black holes is to watch for the shockwaves sent out into space when two of them merge. Because black holes bend spacetime so much, if you have two black holes that come across each other and start to orbit around each other, the space around them doesn't know what's hit it. As the two black holes orbit, how much space bends changes with every orbit. This change sends out ripples through space itself that we call gravitational waves. When the two black holes finally merge, there's a large burst of gravitational waves before space settles back down again. These ripples physically squash and stretch space in tiny amounts as they pass by. We detect this squashing and stretching using extremely sensitive detectors. There are currently two of these detectors, called LIGO, in the United States (one in Livingston, Louisiana, and one near Richland, Washington), and one near Pisa, Italy, called VIRGO. These detectors are made of two mirrors each, which are kilometers apart underground. A laser is bounced between the two mirrors to measure the distance between them with incredible precision, so that we can tell when the space between them gets squashed and stretched by a passing gravitational wave. We can only be

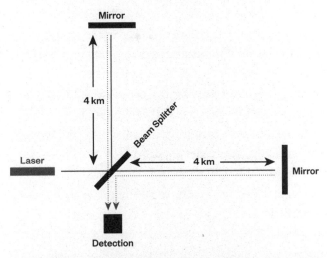

LIGO and VIRGO use lasers and mirrors to detect gravitational waves, which can only be caused by two black holes merging. (Not shown to scale.)

sure that what we detect is a gravitational wave from space if more than one of the detectors, on opposite sides of the continent or planet, detect the ripples at the same time.

Now, black holes are where my heart truly lies—but not regular black holes. I spend my days trying to understand how the supermassive variety of black holes affect the galaxies they live in the center of. These supermassive black holes tend to be about a million to a billion times more massive than the Sun. They're a special class of black holes, separate from stellar-mass black holes, which form when a star goes supernova at the end of its life, leaving a black hole roughly the same mass as other stars. The fact that we humans, such a relatively small chunk of the big bang energy budget, can even hope to understand and study some of the most energetic objects in the entire universe amazes me.

So, how do we find these supermassive black holes at the centers of galaxies? One of the most direct pieces of evidence we have has come from studying the motions of the stars in the very center of our own galaxy over decades. We can see them orbiting around the center at great speeds. We can then use these orbits to calculate how heavy the object they're orbiting around is. Using these methods, we have found that at the center of our Milky Way, in a region of space that could fit inside Mercury's orbit, there is four million times the mass of the Sun. To be that small, and yet that heavy, the only thing that can be at the center of our Milky Way galaxy is a supermassive black hole.

Interestingly, there was a lot of debate in the 1980s as to whether there was a swarm of black holes or a single supermassive black hole in the centers of galaxies. Although a swarm of black holes would have been infinitely cooler, there just isn't enough space in the center of galaxies for a swarm of black holes to be stable. There would be endless collisions and interactions that would quickly disrupt the swarm, sending some slingshotted outward and the rest crashing together in the center to create one massive object.

The evidence we have for supermassive black holes in other galaxies in the universe was, until very recently, not as direct. We could only infer that they were there based on observations of the centers of galaxies billions of light years away. When we observe these we see very high-energy radiation coming from the center; lots of X-rays, radio waves, and high-energy radiation that only occurs when the source of that radiation is also incredibly high energy. You can get this kind of radiation if you've got hydrogen gas being accreted onto a very heavy, compact object—the friction caused by the speed at which the gas is spiraling inward is so great that it causes it to glow in X-rays and radio waves.

After decades observing this incredibly high-energy X-ray and radio emission from the centers of galaxies, astronomers concluded that the only objects that have enough energy to do this are supermassive black holes. This was confirmed by the Event Horizon Telescope collaboration in April 2019 when they revealed the first-ever image of the glowing matter spiraling around the supermassive black hole in the very center of the Messier 87 galaxy. Crucially, this image showed the point at which we do not detect any more light from that material at the event horizon of the black hole—providing the strongest evidence so far that black holes exist and that they behave exactly in the way the theory of general relativity predicts they should.

Along with this emission from this disk of spiraling material, the pressure around a black hole as it tries to accrete that much mass too quickly sometimes means that the least energetic thing it can do is to throw the gas out of this accretion disk into thousands of light-year-long jets, which also emit radio waves. It's this energy that's thrown out from the accretion disk around the black hole as it gets too greedy that we think can affect the galaxy it resides in. For the galaxy to keep on changing and evolving, it needs to keep forming new stars. For that, it needs cold hydrogen gas that can collapse together under gravity until it finally reaches the point where it's dense enough to ignite nuclear fusion. According to our best theory of the universe, the energy thrown off from the accretion disk around the supermassive black hole has to either heat or expel that cold hydrogen gas, which we predict reduces the rate of star formation and stops massive galaxies from growing too big. However, we've never actually observed that happening across a large sample of galaxies—something that I continue to search for in my day job.

To put this all into perspective for Messier 87, the galaxy itself is hundreds of thousands of light years across but the giant jet of radio emission burped up by the supermassive black hole is more than 10 million light years long. If Messier 87 was the size of a grain of sand, the supermassive black hole at the center would be the size of an atom and the jets from it would extend across the entire palm of your hand. That factoid always makes me think of a verse of William Blake's:

> To see your world in a grain of sand, and heaven in a wild flower.
>
> To hold infinity in the palm of your hand, and eternity in an hour.[3]

I doubt that Blake had Messier 87 in mind when he wrote his poem—considering it was 1863 and no one knew then that in the heavens existed billions of galaxies, each with a supermassive black hole at its center, possibly expelling radio jets so unfathomably long it is hard for us to comprehend them. But it's what I think about every time I hear it.

3 Blake, W., and L. Baskin. 1968. *Auguries of Innocence*. New York: Printed anew for Grossman Publishers.

Just Because You Haven't Seen It Doesn't Mean

Doesn't Mean

It Doesn't Exist

4

Every single object you see around you right now, even this very book you're holding, is made out of ordinary matter. That is, they're made of baryons—protons, electrons, and neutrons. All of this sort of matter interacts with light in some way, either producing light, reflecting it, or absorbing it. When that happens, we can detect something is there either by the presence of light or the absence of it. When we look out into the vastness of space, every single thing that we can see is made up of that ordinary matter, from stars, to dust, to black holes. And yet ordinary matter makes up only 15 percent of all the matter in the universe. We don't know what the remaining 85 percent of all the matter in the universe is made of. It's a sobering thought.

We call the rest of the matter that doesn't interact with light "dark matter." It doesn't give out light, or reflect or absorb it, so we have no way of "seeing" it. Unlike an isolated black hole, which traps light beyond its event horizon (we could, perhaps, class that as "absorbing" light), dark matter pervades the whole universe. Even in the solar system, we think there's about two protons' worth of dark matter in every teaspoon of space, and yet we've no way of interacting with it.

So, how do we even know dark matter exists? Well, just like black holes, although light is not an option, dark matter does interact with gravity and so bends space in exactly the same way as normal matter. And, in the last half century, there have been a couple results indicating it has to be there.

The first piece of evidence came from the rotation speeds of stars in galaxies. This concept was being studied back in the 1970s by Vera Rubin, who pioneered the study of dark matter. It's no small thing to be the first person to find evidence for something that makes up 85 percent of all mass in the universe. Ironically, Vera was looking for a problem that she could work on that nobody would bother her about and that would keep her out of trouble (her PhD thesis results had caused some controversy among astronomers at the time—despite later turning out to be right).[1]

Vera thought that measuring the speeds at which the stars were moving around the centers of galaxies at various different distances would be a safe bet.

The law of gravity tells us that because there tend to be more stars in the center of a galaxy, the speed that stars

---

[1]   Rubin had previously looked at the rate of the expansion of the universe and found that it seemed to be different in different directions. This was controversial at the time but was eventually found to be true locally, if not universally—like not being able to see the woods for the trees.

orbit around the center should drop off as you get farther out. This is what we see for planets in our own solar system—the planets closer to the Sun orbit much faster than the planets farther away from the Sun. That's because more than 99.8 percent of the mass in our solar system is concentrated in the center in the Sun. The question that Vera was trying to answer was whether the same thing is found for the much larger scale of stars orbiting within galaxies. So, Vera needed to measure the speed at which stars were moving. This sounds as if it should be complicated but it's quite a neat bit of physics.

Light is a wave (except for when it's a particle—it has a bit of a personality disorder, does light), just like a sound wave. The same process that causes an ambulance siren to become higher pitched by squashing the wave when it's coming toward you, and to become lower pitched by stretching the wave when it's moving away from you, also affects light. This is called a Doppler shift, but instead of a pitch change, squashing and stretching light waves changes the color of light we see. More stretched-out light waves are redder and more squashed light waves are bluer. As we look at a galaxy spiraling around, some of the stars will be rotating toward us and some will be rotating away from us. We see a different Doppler shift in the light from one side of the galaxy to the other. That Doppler shift tells you how fast the stars are moving.

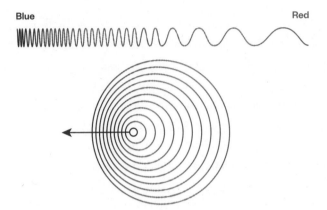

Blue   Red

The Doppler shift. As light waves travel, the color of light we detect shifts blue when objects are moving toward us and red when objects are moving away.

Starting with the Andromeda galaxy, Vera found that the stars were not rotating at slower speeds in the outskirts of galaxies; in fact, the speed was pretty constant all the way out from the center. Very strange indeed and definitely not what was expected. What this finding suggested was that most of the mass in galaxies isn't found in the center—it's found on the outskirts. Yet when we look at a galaxy, we don't see more stars on the outskirts; the greatest concentration of stars occurs in the center. So, that means there must be a huge amount of matter that we can't see surrounding a galaxy, in something we've come to call a *dark matter halo*.

The second piece of evidence for dark matter came from comparing how much mass gravity suggested was in galaxies to how much we could detect in starlight. As Einstein said, massive objects curve the spacetime that light travels along. This might sound like an odd concept, but lenses also change the path of light. Your glasses or contact lenses bend light so that objects, either in the distance or close to you, are focused specifically for your eyes. Massive clusters of galaxies can have the same effect on background galaxies, so that the light path gets curved around the cluster on its way to us. If the alignment is exactly right, you can even get the background galaxy forming a ring of light around the foreground galaxy. In astronomy, we call this an Einstein ring, but you may have noticed that you can recreate a similar ring of light with a stemmed wine glass and a candle flame on a romantic evening.

The amount that the light is bent into a ring tells us how much matter is in the foreground galaxy doing the lensing. Once again, we find that when we see these galactic lenses, gravity says there is more matter there than we can actually see in the light from the stars.

There has been a lot of debate in the astronomical community throughout the past couple of decades about whether we could make up this missing matter with very faint stars, neutron stars, and black holes. We refer to these as MACHOs—**MA**ssive **C**ompact **H**alo **O**bjects. Very faint stars tend to glow in the infrared because they're still hot, so we can estimate how many of those there are. We can

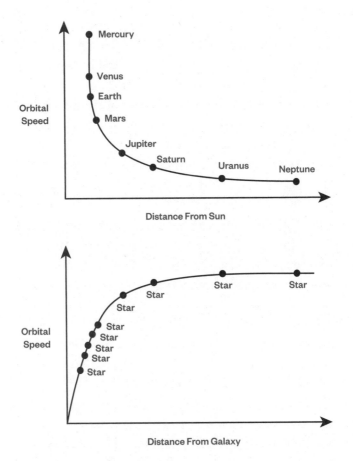

Within the solar system, planets closer to the Sun orbit around it faster than planets farther away (top) because the majority of the mass in the solar system is concentrated in the center. In a galaxy, stars nearest the galactic center orbit slower than stars farther away (bottom). This suggests that most of the mass in a galaxy is found in its outskirts. Since there aren't many stars at that distance, galaxies must contain large masses of dark matter.

detect neutron stars in other wavelengths of light, such as radio waves and X-rays, allowing us to estimate their number. Stellar-mass black holes act as "micro lenses" in our own Milky Way, briefly brightening a star as it passes in front of it, from our perspective. Considering how many of these events we see per year and the chance that such an event lines up from our position on Earth, we can estimate how many of those there may be that we can't see. However, putting all this together and knowing how heavy these objects tend to be still doesn't give us enough matter to explain the amount of it that lensing tells us is there.

Some astronomers think that all these things could be explained if Einstein got gravity wrong. There's a field of research referred to as MOND—**MO**dified **N**ewtonian **D**ynamics, which is the theory that Newton, not Einstein, had it right all along,[2] and rather than needing dark matter, all we have to do is tweak Newton's math ever so slightly to explain what happens at faster speeds and higher masses. But, when it comes to explaining other observations that we've made of huge galaxy clusters and solving this problem of missing matter, these theories fall down. The biggest

---

2   Newton's law of gravity correctly explains everyday objects with low speeds here on Earth, but in their original form, Newton's equations don't allow us to predict the orbit of Mercury properly or the matter spiraling around black holes. Einstein's laws of gravity, on the other hand, work in both the everyday regime and the incredibly fast, extremely heavy regime of objects in space.

issue for MOND right now is that it predicted that the speed at which gravitational waves would move through space would be different from the speed of light. However, that was disproven in 2017 with the simultaneous detection of both gravitational waves and a burst of light coming from a merger of two neutron stars into a black hole. None of the MOND theories has come close to explaining everything that Einstein's theory of general relativity does so elegantly, and so we must accept that dark matter does exist in some form in the universe.

What form dark matter takes is still a question to be solved by particle physicists. They're on the search for WIMPs—**W**eakly **I**nteracting **M**assive **P**articles—as opposed to the MACHOs that astronomers first checked for. Although the particles are termed "massive" by particle physicists, they wouldn't be classed as massive in anyone else's book: they are about a hundred times more massive than a proton (a proton is about one octillionth of a kilogram).

Particle physicists have a truly beautiful theory. It's called the Standard Model of particle physics, and it pieces together all the building blocks of matter—the particles that govern the four main forces (gravity, electromagnetism, and the two atomic forces: the strong force and the weak force)—and explains why things have mass in the first place. It condenses all this into one single equation that describes everything we see in the universe around us.

But it does have some problems—the biggest being the fact that it doesn't incorporate dark matter at all. So, there are now some particle physicists trying to extend their beautiful theory into something a bit clunkier to shoehorn dark matter in there. To do that, they need to find what dark matter is made of—by detecting some of it.

How to search for something that tiny, something incapable of interacting with light or even other matter, for that matter, is a big problem. One method is to attempt to detect when a particle of dark matter collides with a normal particle of matter that we can detect. We should then be able to identify the change in momentum, or energy, of the other particle, like a cue ball hitting a colored ball in a game of pool. Those kinds of collisions are part and parcel of the day-to-day life of a warmish fluid, like water or air. So, physicists first have to supercool a fluid down to a fraction above absolute zero degrees so the particles have no energy and are barely even moving. They do the supercooling miles underground to shield the fluid from other radiation, such as cosmic rays from space, which could cause a bump in a fluid particle's energy. Once you have your supercooled, supershielded fluid miles underground, it becomes a waiting game for the moment you detect a jump in energy caused by an unlikely, but not impossible, collision with a dark matter particle.

These kind of experiments have been running since 1996—and we're still waiting.

How Far We'll Go

5

The farthest distance a human being has ever ventured from the surface of the Earth is 400,171 kilometers (248,655 miles) on April 15, 1970. This record was set during the attempted Moon landing mission of *Apollo 13*, during which an oxygen tank explosion crippled the ship, meaning that a safe landing on the Moon was no longer possible. At the time of the explosion, the astronauts—Jim Lovell, Fred Haise, and Jack Swigert—were more than fifty-five hours into their sixty-hour journey to the Moon. They were too distant from Earth for NASA's original mission abort contingency plans to bring the crew back to Earth safely. Instead, the crew looped around the far side of the Moon to give them a gravitational energy boost to help them get back to Earth. It was during this loop around the far side of the Moon, when the Moon also happened to be at its farthest position from Earth in its orbit, at an altitude of 254 kilometers (about 158 miles) above the surface—more than 100 kilometers (about 62 miles) higher than the flight paths of other Apollo missions—that this record was set.

All three astronauts were safely returned to Earth six days after launch thanks to the ingenuity of the crew and NASA ground staff. It's difficult, then, to say whether this mission should be considered a failure or a success. The original objective of landing on the Moon wasn't achieved, yet no human life was lost and a new human spaceflight record was set. Almost fifty years on, we have achieved new records for the longest time in space, the most space walks, the longest space walk, and even untethered space walks. Yet, all these achievements have been set in orbit around our own planet, either in the now-retired Russian space station, *Mir*, or the currently occupied International Space Station. Since 1972, no human being has ventured farther than 620 kilometers (335 miles; that's just slightly more than the distance between Los Angeles and Las Vegas) from the surface of the Earth.

One reason for that is surely cost. It's cheaper, easier, and more feasible to send a robotic lander or probe to investigate our solar system's backyard. The human race is particularly enamored of Mars. After NASA's Mariner spacecraft flew by Venus in 1963, and we learned that the atmosphere was a smog of acid and carbon dioxide, with a surface temperature of over 400°C (752°F), our focus firmly settled on Mars as the world most like our own.

Discounting our Moon, we have sent as many probes to Mars as we have to all other bodies in the solar system combined. At roughly half the size of Earth, with a year twice as long and a day similar to ours at 24.5 hours long, it's easy to see why people are so curious.[1] Crucially, we all want to know whether Mars does, or could ever, host life.

But one key difference between Earth and Mars could hamper efforts to make Mars habitable for humans: its lack of magnetic field. At first, you might think a lack of magnetic field will only affect navigation on Mars, as compasses won't work without a North Pole to point at, but it has far more serious consequences than that.

The Earth's magnetic field protects us from harmful energetic particles from the Sun. Although the Sun gives off visible and UV light that allows us to thrive on Earth, it also emits across the full spectrum of light—from X-rays to radio waves. The ozone layer at the top of our atmosphere absorbs a lot of the high-energy light, such as X-rays and UV light, preventing it from penetrating down to us on the surface, but the atmosphere is powerless at blocking high-energy particles that the Sun burps up occasionally when the pressure of all those nuclear fusion reactions gets to be too much. This is where Earth's magnetic field steps in.

---

1   Some researchers, engineers, and trainee astronauts live on "Mars time" to support experiments here on Earth. In particular, the long-running HI-SEAS experiment on Mauna Kea, Hawaii, is trying to simulate what a human space base on Mars would be like. Although the researchers quickly get out of sync with Earth time, people often report that living with a day 24.5 hours long is ideal—it gives you that extra half hour to get the little things in life done.

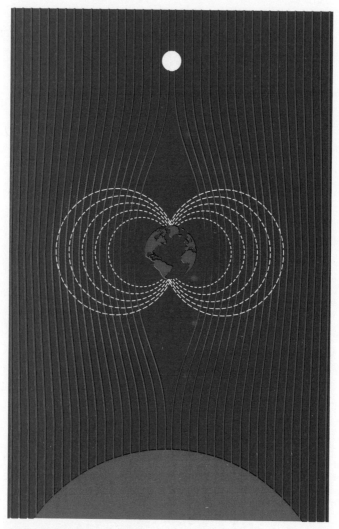

Earth's magnetic field protects it from solar radiation, but Mars has no magnetic field.

It shields the Earth from the majority of the particles in this "wind" from the Sun, whereas the remainder is funneled down to the poles, where they hit the nitrogen and oxygen in our atmosphere and give them enough energy to glow. We call these spectacular light shows the Aurora Borealis and Australis (or the Northern and Southern Lights). We're not the only planet in the solar system that gets treated to these incredible light shows at the poles either; images of Jupiter and Saturn's aurorae are particularly spectacular. But Mars doesn't have such strong aurorae because it doesn't have a magnetic field.

This lack of magnetic field means Mars has no protection from that barrage of high-energy particles from the Sun. Over the past five billion years, since the formation of Mars and our solar system, its atmosphere has been fighting a losing battle against the pressure exerted by these particles. As a result, Mars has been stripped of all but the heaviest elements that it has been able to cling on to in its atmosphere. Its atmosphere is now more than 98 percent carbon dioxide and 170 times thinner than Earth's atmosphere. With no ozone layer to protect it, harmful radiation such as X-rays and UV radiation can easily get to the surface, further complicating life for any adventurous humans who maybe, one day, decide to set up shop on Mars.

Currently on Mars there is one operating rover, *Curiosity*, and one stationary lander, *InSight*. Together with their predecessors, they have explored and investigated more of the surface of Mars than probes on the Moon ever have.[2] The combined efforts of these rovers have allowed us to determine that water probably once flowed on Mars before either being locked in the polar ice caps or, perhaps, evaporating and then being lost to space. We lost contact with the *Opportunity* rover in 2018 after fourteen years of operation on Mars's surface, setting another space exploration record of 45.16 kilometers (28.06 miles) of Mars's surface covered by a single rover.

The reason we lost *Opportunity* is very telling for future human bases that might be built on Mars. *Opportunity* was solar powered. It could survive a couple of days without charging from the Sun provided it didn't roam too far, and it could survive a couple of months without charging from the Sun by going into hibernation mode. Hibernation can happen if there is a particularly large dust storm on Mars. Dust storms are created in much the same way storms are created on Earth. The Sun heats the air closest to the ground. This hot air rises, and as it does, it takes surface dust particles with it, until dust clouds have formed above the surface (unlike on Earth, where water evaporates from the surface to form clouds high in the atmosphere). On occasion, when

---

2   The lunar rovers driven by *Apollo 15, 16,* and *17* astronauts covered between 27 and 36 kilometers (16.7–22.4 miles) each, tipping the full distance traveled on the Moon over that traveled on Mars by robotic rovers.

the conditions are right and Mars is close to the Sun in its orbit, these dust storms can grow to engulf the entire planet and block out nearly all light from the Sun. At such times, a rover needs to be able to turn off all systems and conserve power. Periodically, a rover is supposed to wake up, check its battery situation, and, if it has enough power, send a ping to Earth. *Opportunity* went into hibernation mode in June 2018 and the dust storm abated in October 2018; however, by February 2019, NASA still hadn't received any communication from *Opportunity*. By that time, any dust accumulated on the solar panels should have been blown away, so it is presumed that dust must have somehow gotten into the electronics and finally ended *Opportunity's* mission.

These planet-wide dust storms are expected, roughly, every three years, so they're not something you can ignore when planning for a base on Mars. You can imagine that, faced with such a storm, a lack of power for months on end is going to be a real issue for a human base. Not only for communications and Netflix binges, but also for crucial life-support systems such as oxygen regulation and heating or cooling systems. Before we can even think of setting up a human base on Mars we'd need an airtight dust storm contingency plan, which may involve a mission abort in the worst-case scenario.

Future destinations in our solar system neighborhood include potential probe missions to a few moons of Jupiter,

Saturn, and Neptune—mainly by virtue of them being possible candidates for life, with their large oceans buried beneath icy crusts, plus intense volcanic activity. But getting humans to explore these possibly habitable worlds is a big issue in space travel. The record for the fastest-ever human spaceflight was set by the *Apollo 10* crew as they gravitationally slingshotted around the Moon on their way back to Earth in May 1969. They hit a top speed of 39,897 kilometers per hour (24,791 miles per hour); at that speed you could make it from New York to Sydney and back in under one hour. Although that sounds fast, we've since recorded un-crewed space probes reaching much higher speeds, with the crown currently held by NASA's *Juno* probe, which, when it entered orbit around Jupiter, was traveling at 266,000 kilometers per hour (165,000 miles per hour). To put this into perspective, it took the *Apollo 10* mission four days to reach the Moon; *Opportunity* took eight months to get to Mars; and *Juno* took five years to reach Jupiter. The distances in our solar system with our current spaceflight technology make planning for long-term crewed exploration missions extremely difficult.

So, will we ever explore beyond the edge of the solar system itself? The NASA *Voyager 1* and *2* spacecraft were launched back in 1977 with extended flyby missions to the outer gas giant planets of Jupiter and Saturn. *Voyager 2* even had flyby encounters with Uranus and Neptune—it's the only probe ever to have visited these two planets. The detailed images you see of Uranus and Neptune were all taken by *Voyager 2*. Its final flyby of Neptune was in October 1989, and since then, it has been traveling ever farther from the Sun, to the far reaches of the solar system, communicating the properties of the space around it with Earth the entire time. In February 2019, *Voyager 2* reported a massive drop off in the number of solar wind particles it was detecting and a huge jump in cosmic ray particles from outer space. At that point, it had finally left the solar system, forty-one years and five months after being launched from Earth.

Voyager 1 was the first craft to leave the solar system in August 2012, and it is now the most distant synthetic object from Earth at roughly 21.5 billion kilometers (13.5 billion miles) away. *Voyager 2* is ever so slightly closer to us at 18 billion kilometers (11 billion miles) away. Although we may ultimately lose contact with the Voyager probes, they will continue to move ever farther away from the Sun with nothing to slow them down or impede them. For this reason, both Voyager crafts carry a recording of sounds from Earth, including greetings in fifty-five different languages, music styles from around the world, and sounds from nature—just in case intelligent life forms happen upon the probes in the far distant future when the future of humanity is unknown.

The next nearest star to the Sun is α-Centauri at an epic 39,923,400,000,000 kilometers (that's almost 25 trillion miles) away.

It takes light a little over four years to travel that distance at 299,792 kilometers per second (186,282 miles per second). At the more realistic speed of *Voyager 1*, it would take the probe around seventy-four thousand years to reach α-Centauri, except that it's not heading anywhere near there. Instead, it's heading in the direction of the constellation of Ophiuchus, so that in around forty thousand years, *Voyager 1* will come within 16 billion kilometers (10 billion miles) of a star in the constellation of Ursa Minor, and its closest star will no longer be the Sun, which gave life to all the recordings it carries.

Unless we manage to come up with another, more efficient, and faster way of powering spacecraft, it's not going to be an easy task to get a human being to abandon their friends and family by signing up for a one-way trip to the far reaches of the solar system and beyond. If we ever do want to explore beyond the safety of our Sun's gravitational sway with interstellar travel, we're going to need better technology and some extremely intrepid explorers. I wonder, reader, if you were given the chance to go today, would you?

Apollo 13 used gravity to slingshot around the Moon and return to Earth.

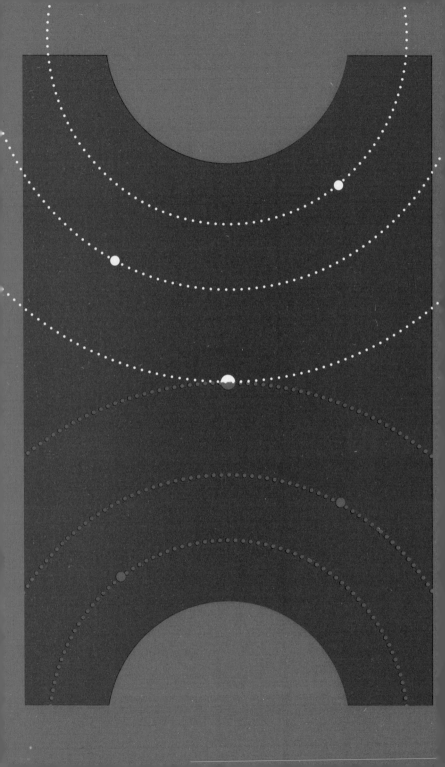

SIX

In Pursuit of Earth 2.0

The universe sprang into being, in a wondrous storm of creation, 14 billion years ago; 9.5 billion years later, the Sun, the Earth, and the solar system started to take shape. After another billion years, life developed in the Earth's oceans, paving the way for dinosaurs to walk the land (a mere 13.5 billion years since the big bang). Fast-forward through the next 500 million years, with the evolution of plants, mammals, and birds, through the rise and fall of the Greek, Roman, and Mayan civilizations, until we arrive in 1995. The Spice Girls are at the top of their game, *Toy Story* has just come out in the cinema, and— finally—the first confirmed planet around another star in the Milky Way is found. The discovery of this planet, dubbed the very lyrical sounding 51 Pegasi b, is, for me, one of the greatest astronomical discoveries to occur in my lifetime.[1]

This discovery confirmed that our solar system was not merely an oddity in the universe—that other stars also had planets orbiting them and that maybe, just maybe, those planets might host life.

---

[1] Although, admittedly, it's a discovery that passed me by at the time because I was too focused on the new Spice Girls album.

Furthermore, it kicked off what has been referred to as the "golden age" of exoplanet research, where discovering a new planet has become the norm for a Tuesday afternoon. It's something any keen amateur astronomer could get involved with if they wanted. But the real question, even before that first confirmed discovery in 1995, has always been, where is the most Earth-like planet? We want to know if there are other planets out there like ours, orbiting stars like ours, and possibly hosting life like ours. In recent years this quest has shifted from mere curiosity to one that feels as if it's driven by necessity, even though—as discussed in the previous chapter—moving to another Earth is unlikely to be a viable option for us, no matter how much we might end up needing one.

The impending doom of our own planet aside, how do we go about finding an extrasolar planet? There are three main methods people use to find these exoplanets: direct imaging, radial velocity measurements, and stellar transits. The first, direct imaging, does exactly what it says on the tin. You take a direct image of the planet orbiting around its star. While this might sound easy, remember, stars shine and give out a rather large amount of energy in order for us to be able to see them so many millions of miles away. Planets, however, do not shine; they merely reflect the light from their stars so we can see them. This is how we see the Moon and planets in the solar system.

Of course, that means that the star is significantly brighter than the planet when you try to take an image, and this makes the planet incredibly difficult to spot—like trying to see someone holding an LED next to the glare of a stadium floodlight.

The trick to this method's success is, first, to block the light from the star, essentially by placing a circular mask over the middle of your telescope detector, and then take an image. As there's lots of noise in those images, people tend to wait a couple of days, or weeks, before taking another couple of images. This wait is so we can check that the bright collection of pixels we thought was a planet in the first image is definitely still there in the second image (rather than just being a speck of noise). It may even have moved further along its orbit around its star in that time, too.

Planets that orbit farther away from their star are easier to spot in a direct image, as it's easier to separate their reflected light from the glare from the star. I'm talking about planets that are more than a hundred times farther away from their star than Earth is from the Sun. For context, Pluto is only fifty times farther away from the Sun than Earth is. So, if we only used this method to find planets, we'd bias ourselves to always finding more massive planets, which reflect more light and orbit at great distances from their star. Not exactly the best method for finding Earth 2.0.

Another possible method is radial velocity measurement, which relies on the fact that the center of mass of a planet-star system is not precisely in the middle of a star. It's easy to think of the Sun as the very center of Earth's orbit but, in reality, that's not the case. If you imagine two objects of the same size and mass orbiting around each other, they'll orbit around the point exactly halfway between them, which we call the center of mass. If we make one of those objects steadily more massive, it will have a greater gravitational pull than the other and the center of mass will shift toward the heavier object. In the solar system, the Sun is so massive that the center of mass has shifted to the point that it's somewhere inside the Sun but not perfectly in the center. Because the Sun continues to orbit around the center of mass, it appears to wobble. In fact, all stars with planets have such a wobble. Thus, at some points in a planet's orbit, its star is wobbling toward us, and at some points it's wobbling away from us. This wobbling stretches and squashes the light from the star to redder and bluer colors (the same method Hubble used to work out that galaxies were all moving away from us); by detecting this wobble, we can calculate the velocity shift of the star.

Using some simple orbital mathematics, we can then relate this velocity shift to the mass and orbiting distance of the exoplanet around the star. Since the planet is orbiting the star, we'll see this signature velocity change

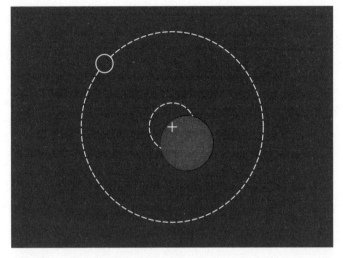

The center of mass of a planet-star system is not perfectly in the center of the star.

repeating, so we can be completely certain when we've discovered a planet. Again though, this method comes with a fair few biases. The bigger the planet, the bigger the velocity shift it will cause its star to have. On top of that, the repeating signature is much easier to detect for planets that are closer to their star, with rapid orbits. If we were looking for an Earth-like planet that took a full year to go around its star, we'd have to observe the star for more than a two-year period to see that repeating velocity shift. In comparison, a planet that only takes a couple of days or weeks to go around its star is much easier to spot.

Back in 1995, radial velocity measurements were used to detect 51 Pegasi b, a roughly Jupiter-size planet orbiting around a Sun-like star every four days. It's unsurprising that this was the first planet we detected as it's the Goldilocks of exoplanets: the star is not too massive, the planet is not too small, and its orbit is not too long. All this, combined with the sensitivity of our telescopes back in 1995, meant the conditions to detect 51 Pegasi b were ripe. Its presence causes its star to have a velocity shift of 70 meters (76.5 yards) per second. For comparison, Jupiter causes the Sun to wobble about 13 meters (14 yards) per second, and the Earth only causes a 9-centimeter (3.5 inch)-per-second wobble. That kind of shift is tiny in comparison to the one caused by 51 Pegasi b, a sensitivity our detectors are only reaching now.

By far the most successful method of detecting exoplanets has been using the transit method. This is when a planet passes in front of its star and causes a dip in the star's observed brightness. Again, if we wait long enough, we may also see this dip in the brightness repeating as the planet passes by the star on each orbit. All we have to do is stare at a patch of sky and continuously record the brightness of all the stars in that patch, then sort through all the data to find the dips in brightness. This is exactly what the Kepler space telescope did from 2009 to 2018. The science team even enlisted the help of the public to find those

A planet passing in front of a star will temporarily reduce the star's observable brightness.

characteristic dips in brightness that a computer searching through the data might have missed. By doing so, citizen scientists managed to find a seven-planet system around a star, a weird-looking signal that the computer wasn't programmed to find. Over the years this transit method has been used to find more than 7,500 exoplanets (compare that number with the 456 found from radial velocity measurements and a mere 19 via direct imaging).

Again, this method doesn't come without its biases, meaning those 7,500 exoplanets give us a skewed picture of planets in our Milky Way. The bigger the planet, the bigger the dip in brightness produced, hence the first planet detected with this method caused a 1.7 percent dip in brightness of its star. An Earth-like planet around a Sun-like star would cause only a meagre 0.008 percent dip in brightness. What's more, the closer the planet is to the star, the more repeat dips we see and the more confident we can be of a detection. As a result, we tend to find large planets close to their stars with this method.

Combining all these methods, we have an overabundance of hot (i.e., close to their star) Neptune- and Jupiter-like planets rather than Earth-like planets. Having said that, our detectors are getting more sensitive and exoplanet missions are working long term, meaning we do have a fair few candidates for the title "most Earth-like planet"—but first we must define what that title actually means. It could be the planet most similar in size to Earth, or most similar in mass, or with an orbit that takes about a year, or one that's at the same distance from its star as Earth. If, however, you're looking for a planet that would allow life to thrive, then, unfortunately, most of the exoplanets we currently know of that are closest in size or mass to Earth tend to be too close to their star to allow for this.

However, not all hope is lost. There is also Kepler-438b. This is incredibly similar in size to Earth, but it orbits much closer to its star so that its year is only thirty-five days long.

The good thing is that its star is much cooler and smaller than the Sun, so that the average temperature on the surface of Kepler-438b is about 3°C (37.4°F). Now that we could definitely work with, if we do ever wish to jump ship from Earth. The problem is that Kepler-438b is 640 light years away. A light year is the distance covered by light, traveling at about 300,000 kilometers per second, in a single year. That's about 3,000 trillion kilometers away. Even at the speed of light, it would take us more than half a millennium to get there, and as we've heard, current technology isn't anywhere close to that speed. So, even if we had the capability to travel there, not even your great-great-grandchildren would be alive to see the ship arrive. And who knows what we might find if we do make it there: perhaps a whole host of different species that already call Kepler-438b home.

Why the Night Sky
Is Dark

7

The question about why the night sky is dark has been posed by many physicists and philosophers over the millennia—from the ancient Greeks through to twentieth-century astronomers. It was popularized in the nineteenth century by a German called Heinrich Olbers—a doctor by day and an astronomer by night. Although many others had described this problem before, Olbers formulated an explanation that was named after him: Olbers's paradox, sometimes known as the dark sky paradox. You might think there is a simple answer to this question: surely, the night sky is dark because the Sun has set? But, as the Earth spins on its axis to face us away from that great big life-giving ball of light that is ever present in our daytime sky, it turns toward uncountable numbers of other stars. Although these stars may be farther away, there are a great many more of them—enough to make our home star seem rather insignificant. The answer is, therefore, rather more profound, giving us insight into the very nature of the universe we live in.

Think back to what previous generations have considered to be true about the universe. In the sky are the Sun, the Moon, the planets, and the stars that would always rise again after setting. These things were known and they were a constant. Based on these observations, our ancestors drew these conclusions about the universe:

1. That it was the same in all directions because you see stars in every direction you look (we call this a homogeneous universe)

2. That it was unchanging, forever remaining the same, because nothing changed with each passing year (a static universe)

3. That the universe was infinite because, as telescopes grew with the centuries, an ever-growing number of fainter stars were found in every part of the sky

If all these things about the universe are true, then every line of sight, every single place you look in space, should eventually happen upon a star. Imagine taking a small patch of sky, perhaps the size of your thumbnail, which, held at arm's length, is about the size of the Full Moon. Try it next time you're looking at the night sky and you'll find you'll be able to block the Moon out of the sky with your thumb. We are both well aware that, in reality, the Moon is a lot bigger than your thumb. In fact, if I assume my thumbnail is roughly circular and about 1.5 centimeters (½ inch) across, I could fit 50,000,000,000,000,000 of my thumbnails on the surface of the Moon! The reason my single thumbnail can block out the whole Moon—an arm's length away from me down on Earth—is all down to perspective. If we imagine that my arm could grow to twice its length, then at twice the distance away it would take four of my thumbnails to block out the whole Full Moon. Twice the distance again and it would take sixteen thumbnails.

Now, if extending arms wasn't enough of a stretch for you, imagine that my thumbnail also glows. The closer it is to me, the brighter it will appear. The farther away it is, the fainter it will be. This effect is something we're all familiar with from crossing roads at night—judging how far away a car is by how bright its headlights appear. Astronomers have known for a long time how much fainter things get with distance and, as with perspective, brightness depends on the square of the distance. So, if we go twice as far away, things get four times fainter, because $2^2 = 2 \times 2 = 4$. Three times farther away, things get nine times fainter. Ten times farther away equals a hundred times fainter.

So, if we go back to my glowing thumbnail on my extendable arm—at double the length of my arm, I'm going to need four glowing thumbnails to block out the Moon, but they're going to be four times fainter. So, the two effects cancel each other out to give you four thumbnails as bright as the thumbnail only one arm's length from you. Now, imagine taking this to a hundred times my arm length, where we'd have ten thousand thumbnails, which would be ten thousand times fainter and yet still as bright as a single thumbnail at one arm's length—and so on until infinity.

Glowing thumbnails might be a slightly tenuous analogy but I rather like it. Because imagine that instead of a glowing thumbnail we've got a star, and instead of an arm's length we've got a light year. Even with these much larger objects and distances, the same truths hold as with the glowing thumbnail. If I have one star in a patch of sky at one light year away, I would have four in the same patch at twice the distance, which would be four times fainter.

Now imagine this across the whole sky. A sky that is homogenous, the same in all directions, and infinite. At every single light year step, in any direction we look, we'd have the same amount of brightness as a single star only a light year away. And with an infinite number of light year steps, the night sky would be blindingly bright! So, why is the sky dark? Edgar Allan Poe, of all people, once touched on this in one of his many essays:

> Were the succession of stars endless, then the background of the sky would present us a uniform luminosity like that displayed by the Galaxy—*since there could be absolutely no point, in all that background, at which would not exist a star*. The only mode, therefore, in which, under such a state of affairs, we could comprehend the *voids* which our telescopes find in innumerable directions, would be by supposing the distance of the invisible background so immense that no ray from it has yet been able to reach us at all.[1]

---

1   E. A. Poe (1848). *Eureka: A Prose Poem*. Geo. P. Putnam, New York.

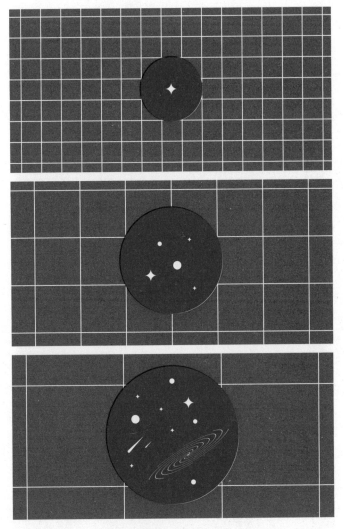

The more sensitive a telescope is, the fainter and more distant the objects it can detect.

Edgar focused on only one reason the sky might be dark: because the universe isn't infinitely old. It has a certain age, in years, since its creation. Adding that assumption to our three previous ones—and taking into account that it takes light time to get to us, traveling at its set speed limit—it follows that we can only see the light of the stars that has had enough time to get to us since the beginning of creation. We now know that the universe was created in the big bang 13.8 billion years ago, so because light will take time to reach us, there will be an "observable universe" beyond which we are blind and we can't see any stars.

But the other thing we know from the theory of the big bang is that the universe is not infinite. It has a finite size. So, not only will light take time to reach us, but we won't have an infinite number of light year steps along a single line of sight. As a result, we won't have the infinite number of stars needed to make the night sky bright.

At the same time, we must remember that the universe is always growing, because space itself is expanding. This expanding of space stretches the light waves traveling across it. The farther light travels in the universe, the more it gets redshifted, just like the stars wobbling away from us and Hubble's galaxies from our earlier discussion. But space has expanded so much that all the visible light from the most distant things in the universe has been stretched beyond visible red light into infrared and even microwaves. These waves are no longer visible to our feeble human eyes, and so, the true brilliance of the night sky is, in reality, entirely hidden from us.

Aliens Probably Exist

When people find out you're an astronomer, they want to ask you about aliens—specifically, whether intelligent life forms similar to humans exist elsewhere in the vastness of the universe. I would argue that this is more a question for philosophers and that studying the effects of black holes on galaxies doesn't give me automatic clout to comment on one of the biggest questions ever posed by humanity. But when people *do* ask me, this is how I answer.

We are one planet, orbiting around one star, in a galaxy of over 100 billion stars. Therefore, it might seem reasonable to assume that one in 100 billion stars can host life. That's not necessarily true, however, because our Sun is an average star. Not too hot, not too cold. Not too energetic, not too much variability in how much energy it produces. In terms of stars it's pretty small, and also very long-lived. Not all stars are like the Sun: the bigger the star, the faster they burn through their fuel, whereas the smallest stars burn through their smaller amount of hydrogen fuel much more slowly. It follows that, to develop intelligent life similar to humans on a planet orbiting a different star, you're going to need a star that will live as long as the Sun.

Before we can determine whether these conditions exist elsewhere in the universe, we first need to establish how old the Sun is and how long it took for intelligent life to develop on Earth. And in order to do that, we need to know how much energy the Sun has in total. The way that stars fuel themselves is by converting four hydrogen atoms (each with one proton) into helium atoms (with two protons and two neutrons) in a process called nuclear fusion. But one helium atom is $7/1,000$ times lighter than four hydrogen atoms. So, where does that missing mass go?

You may have heard of Einstein's most famous equation, $E=mc^2$. The E here is energy, the m stands for mass, and the c is that pesky speed of light. What this equation so beautifully underpins is one of the most fundamental principles of physics: that energy and mass are essentially the same thing. So, the difference in mass between the four hydrogen atoms and the helium atom isn't "missing" at all—it gets converted into energy. It's this energy that we, on Earth, are so thankful for because it provides warmth that has allowed life to thrive, food to grow, and many trips to the seaside.

If we measure how big the Sun is, knowing how dense hydrogen is, we can then work out how much the Sun weighs and how much energy it has. Firstly, note that the whole Sun won't actually get converted into helium because only about 10 percent of the inner part of the Sun is hot enough to force hydrogen atoms together to make helium. So, we need to work out how much mass the Sun has in its center that it can convert to energy and compare that with the Sun's brightness, or how much energy the Sun gives off per second. We can then work out how long the fuel in the Sun will last in seconds. Combining all those measurements together tells us the Sun will live for about ten billion years.

Next, using radioactive age dating on asteroids that fall to Earth, we can work out how long the Sun has been burning its fuel since it formed. These asteroids are the relics of the early solar system when the Sun first formed. The oldest are about five billion years old, so we can assume that the Sun is about halfway through its lifetime. We think that the planets took about 100 million years to form and then another 500 million years for life (at least very simple life, like bacteria) to develop on proto-Earth. But to develop life intelligent enough to leave our own planet, and contemplate this question of whether we could ever find extraterrestrial life, we've needed that full five billion years.

So, that rules out all the stars above two-ish times the mass of the Sun, as they'll run out of fuel far too quickly for life to develop. If you had a star that was smaller and cooler than the Sun, the orbiting planet would need to be closer to the star in order to achieve temperatures fit for life. But when a planet is closer to the star, there's a danger of its atmosphere boiling off, or even its orbit becoming tidally locked. Take Venus, for example. Its day is twice as long as its year. So, for an entire year, you have one side of the planet scorching under the intense heat of the Sun and another side freezing in endless night.

So, perhaps it's not one in 100 billion stars that can hold life, but more like one in a trillion stars that are the right size to burn for a long enough time and host a planet in precisely the right location. However, we also have to take into account the Sun's place in the Milky Way galaxy. We're out on the edge, on a spiral arm of the Milky Way. Not too close to the edge and intergalactic space, not too close to the dense center where our supermassive black hole resides, generating lots of high-energy radiation. Either way, radiation would kill off any life on planets around stars not in this galactic "Goldilocks zone."

In addition, for intelligent life to survive, the planet must have the molecular building blocks for life, such as carbon chains, water, and amino acids. All these contain elements like carbon, oxygen, and nitrogen, which were made in the big forges of the universe—those stars more massive than the Sun that burned bright and fast, eventually exploding outward in a process we call a supernova. Runaway fusion during the supernova converts three atoms of helium into carbon, four atoms of helium into oxygen, and so on all the way to iron, with all of these elements thrown off into space. So, the star must form in a part of the universe where there are remnants of a dead star, so that the elements like carbon, oxygen, and nitrogen that are needed for life will be present. So, perhaps, that brings us to about one in a quadrillion.

Once a star has formed, and a planet has formed from those life-giving elements, it then needs to orbit in the zone around the star where it's not too hot and not too cold. Not only that, the planet needs to stay there. That might sound obvious, but when we simulate planets forming from the mess around stars, they tend to like to move around and migrate inward toward their stars. This nicely explains how all those hot Jupiter-like exoplanets we've been detecting have formed around their stars. In fact, considering that these hot Jupiters seem to be quite prolific throughout the rest of the Milky Way, it's quite odd that Jupiter is where it is in our own solar system, and not closer to the Sun than Earth. It is only through its interaction with Saturn that it's been stopped from migrating inward toward the Sun and disrupting all the other planets as it goes. If that had happened, Earth's orbit could have been disrupted, either moving us out of the precious habitable zone around the Sun, or even slingshotting us entirely out of the solar system. So, if we're conservative with our odds of that not happening, does that bring us to one in a quintillion?

When we think about such a large number (a "1" with 18 zeros after it), it suggests that the chance of there being another planet that can host life in our own galaxy of "only" 100 billion stars is very slim. But the Milky Way is not the only galaxy in the universe. When we look out into the sky, we see these islands of billions of stars everywhere we look—and in all different shapes and sizes: spirals, blobs, train wrecks, even penguin-shaped. It is nigh on physically impossible to count the number of galaxies in the universe because a) there are so many and b) how can we be sure we've found them all?

But, to put some sort of lower limit on this, we can use an image that the famous Hubble Space Telescope (HST) took about a decade ago. Astronomers decided to use HST to stare at the darkest patch of sky that we know, in the constellation Fornax, in the Southern Hemisphere night sky, to see what they could find. They took an image that was a 2 × 2 arcminute square patch of the sky. An arcminute is a funny unit—it is a sixtieth ($\frac{1}{60}$) of a degree, and an arcsecond is a sixtieth ($\frac{1}{60}$) of an arcminute. Given that the whole sky is 360 degrees around, it's a very small patch. It's an image that is 5 percent of the size of the Full Moon in the sky. Astronomers didn't really know what to expect to find in this tiny dark patch of sky, but the latest count on the number of stars found in the image of that patch is four. And the latest count of the number of galaxies is about five thousand—everything from beautiful nearby spiral galaxies to distant galaxies that we detect as just a single pixel.

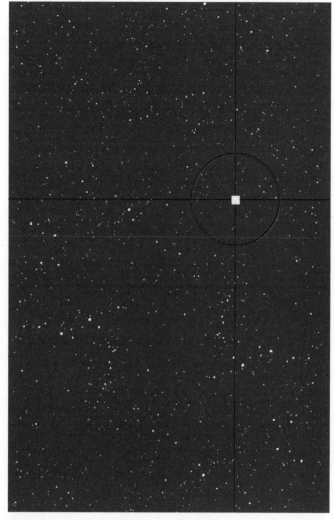

The white square represents a 2 x 2 arcminute square patch of sky relative to the size of the Full Moon (the red circle).

If we take that number and apply it to the rest of the area of the sky, we can estimate that there are at least 100 billion galaxies in the universe. Remember, this image was taken of the darkest patch of the sky, so in other regions we should see even more galaxies (plus all the galaxies that are still too faint for us to see). It's more likely that there are another couple of zeros on the end of that number. So, let's make our estimate a round trillion galaxies. Not only that, let's say that each galaxy contains, roughly, 100 billion stars. So, perhaps, we can estimate that there are at least 100 sextillion stars. That's 100,000,000,000,000,000,000,000 stars in the universe. So, if one in a quintillion stars might develop life, and there are at least a hundred sextillion stars in the universe, then perhaps there are a hundred thousand planets out there in the vastness of space that might have the right conditions to develop intelligent life!

People often ask me, how—as an astrophysicist knowing and thinking about these numbers all the time—I don't get overwhelmed by it all. How do I stare at the sky without being entirely crippled by anxiety at the sheer scale of the whole thing and our own insignificance? Firstly, in day-to-day life—whether you're at your desk crunching data, in an office, at home, or on a train—there isn't time to stop and think about it. But, when I look at the majesty of the night sky, with the Milky Way stretching out overhead in a huge arc of stars, I don't feel anxious. I feel limitless. Like there are infinite possibilities out there and I could be part of any one of them. The scale doesn't scare me; it thrills me. Like the protagonist at the beginning of a good adventure novel yearning to see the world and get out of their small town. When I look up at the sky and think about the sheer number of stars out there, I can't help but get excited about drawing the conclusion that we can't be the only planet whose cards came up right in the game of life.

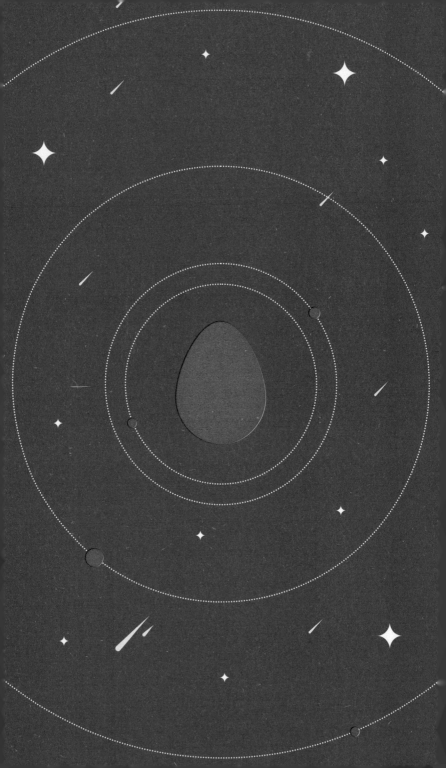

The Original

"Chicken
or the

Egg"

What came first—the chicken or the egg? Darwin's theory of evolution allowed us to answer that question definitively: the egg came first—created by two birds that weren't yet chickens. But astronomy has a far more exciting question: what came first—the galaxy or the black hole?

As I've said, we think there is a supermassive black hole at the center of every galaxy in the gravitational driving seat of the whole star system. So, did the black hole form first, and then the galaxy of stars formed around that gravitational sink? Or did the galaxy of stars form first, with one star going supernova and forming a small black hole that grew and grew until it sank to the center because it had become the heaviest object in the galaxy?

That might seem like an impossible question to answer, as this would have happened in the very early universe after the big bang. But, we can observe the galaxies in the early universe and the clues left over from the big bang to try to figure this out. Right after the big bang, the universe, and space itself, was much smaller, so all matter was condensed into a much smaller volume. It was, therefore, much denser and more energetic, with collisions between particles

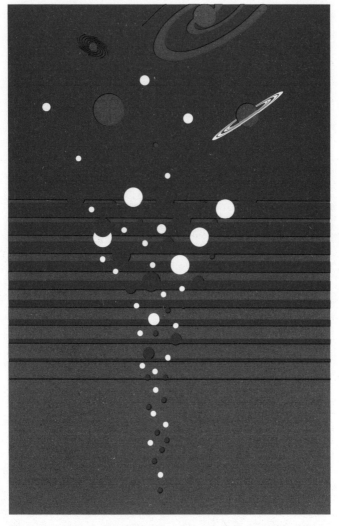

The largest objects in the universe formed from the superheated particle soup created by the big bang.

taking place in small confines. This means that this big soup of particles created in the big bang was also very hot. So hot, in fact, that protons didn't exist until a whole second after the big bang. Although that might not sound like a very long time, within it, a lot happened.

Before the formation of protons from the even smaller quark particles, the universe inflated rapidly so that in a billion septillionth of a second, the universe expanded to 100 septillionth of its size. It's a phenomenon we call *inflation*. It's the highest rate of expansion our universe has ever experienced, and we should be incredibly thankful for it because it meant that the first stars and galaxies could form. Before inflation, tiny numbers of normal and dark matter particles would clump together under gravity, which meant that there were some areas that had slightly more particles in them and some with fewer particles. In this period of inflation, these areas of slightly higher density and lower density were imprinted across the whole universe. It's one main reason our universe appears similar in every direction we look.

By the time the universe had cooled enough for protons and neutrons to form, and once space had expanded enough for hydrogen and helium atoms to form by binding electrons into their orbits, about 380,000 years had passed since the big bang. Because of inflation, more elements, such as hydrogen and helium, formed in the more dense areas, and fewer in the less dense areas. This meant that these newly formed large clouds of hydrogen gas could start to cool and collapse under gravity to form the very first stars, a process we estimate took up to 150 million years.

These areas of higher density meant more stars were forming in certain regions of space, so that they grouped together under gravity, forming the first galaxies.

But how do we then get a supermassive black hole at the center of these newly formed galaxies? Well, we could wait for a giant star to go supernova. Perhaps a star that is a hundred times heavier than the Sun runs out of fuel, explodes, throws out its material into space, and the left-over core collapses down under its own gravity into an eventual black hole about ten times the mass of the Sun. That black hole can then accrete the material thrown out in the supernova to grow bigger.

There could also have been two smaller stars that went supernova—but that weren't quite big enough for their cores to collapse into a black hole. In that situation they'd both collapse into a neutron star, which is the densest form matter can have (that we know of): neutrons arrayed perfectly in a crystal as tightly packed as they can be. If two of these neutron stars merged, you'd get a black hole. Once again, though, we first need stars to form, live, and die before we get to black hole territory. And even then, the black holes that form would be only ten or so times heavier than the Sun.[1] They would have to grow from there to reach supermassive proportions when they're at least a million times the mass of the Sun.

You might think that, if the matter's there, they should be able to accrete it all at once, but that's not physically possible. The accretion rate is limited to something we call the Eddington accretion limit. This limit happens because as material spirals around a black hole, the pressure increases around it as the material heats up due to friction and starts to glow. The light given off is very high energy, so when its impact is felt on other materials, the effect is like a strong wind that pushes the material away from the black hole. This stops the black hole from accreting too much material. It means that, at the maximum possible rate of accretion, it takes about 800 million years for a black hole to grow to 800 million times the mass of the Sun. This is the mass of the most distant growing supermassive black hole we've ever seen—and the light we detected from it was emitted 800 million years after the big bang.

---

1   Stars themselves are limited in how massive they can get because they can only burn so much fuel per second to counteract the constant pull of gravity inward. The heavier the star, the faster it has to burn fuel to counteract the higher gravity. This means that the eventual black hole that can form from the star's leftover core will also have a limited mass when it first forms.

Although a supermassive black hole can form this way, by waiting for a supernova and then growing at the maximum rate for as long as possible, it took 150 million years for our newly formed universe to cool down after the big bang to the point where stars could form. Yet for a star to go supernova, it has to live and then run out of fuel, which takes another 10 million years or so—which means we've got to make our supermassive black hole in 640 million years, not 800 million years. Moreover, that estimate assumes the black hole would be accreting at the maximum rate the whole time, which is highly unlikely. As black holes accrete more material, the hotter that material gets and the greater the pressure pushing material away from the black hole. So, when it starts accreting at the maximum rate, the black hole effectively shoots itself in the foot. What you end up with is sporadic accretion: a period of maximum accretion followed by a quiet period where the gas around the black hole cools down enough to kick-start accretion again.

Alternatively, a black hole could grow by merging with other black holes. Remember, we've detected gravitational waves from mergers of black holes in our own Milky Way, so we know this is definitely possible. We can calculate how many mergers are needed by assuming that the black hole always merges with another black hole of the same mass. The black hole would effectively double its mass with every merger. However, the number of mergers needed to reach supermassive levels would be too high in too short of a time. If you did throw together that many black holes that quickly, you'd end up with a huge swarm of black holes orbiting around each other at huge speeds, and this would disrupt each black hole's orbit, slingshotting some out of the swarm so they'd never end up merging. The gravitational forces in this merging process would also disrupt the material the black hole could accrete, ruling out the possibility that accretion and merging could combine to produce a supermassive black hole. This problem is still puzzling astrophysicists: how did supermassive black holes grow so big, so quickly in the early universe?

There's another theory that's been proposed but not everyone is convinced of it yet. The theory is that you can have the direct collapse of black holes ten thousand times more massive than the Sun from the big clouds of hydrogen gas in the early universe. Imagine that there are two of these large gas clouds next to each other in the early universe and one of them manages to cool, collapse under gravity, and start forming stars before the other. The energy and light given out by those stars will heat the gas in the nearby gas cloud and prevent it from cooling to form stars. But it's also one of the densest parts of the universe, so it's going to attract more gas and dark matter to it from elsewhere, making it even heavier. All the while it's getting heavier, it's still not cold enough for the particles in the gas to stick around in a single place long enough to collapse and form a star. Eventually, the amount of material—both normal and dark matter—in the cloud gets too large and the whole thing collapses under gravity to form a black hole.

This is then the heaviest thing in that denser region of the universe around which stars will form and orbit, ultimately forming a galaxy. In this scenario, the black hole forms first and is at the center of the galaxy from the beginning, rather than the galaxy forming first and an eventual black hole sinking to the center. Similar to our answer for the chicken and the egg, it means that the black hole came first, created by two objects that weren't yet galaxies.

When theorists simulate the early universe on computers they find that, if they include this process of direct collapse black hole formation, it allows them to achieve a good match for our observations of the most distant galaxies and their growing supermassive black holes. Although this has the theorists convinced, I'm an observer; I want to know if we have a chance to observe this collapse happening. Fortunately, some astronomers think they have found a pretty good candidate. It's an object that was found with the Hubble Space Telescope, dubbed Cosmic Redshift 7 (CR7), which is currently 28 billion light years away from us (taking into account the fact that the universe has expanded since the light was emitted). It means that the light from this object was emitted when the universe was only 800 million years old. When we split the light from the object through a prism to get the spectrum showing the signatures of elements in the object, we find a lot of high-energy emission from hydrogen, but barely any of the features we associate with stars. The hydrogen emission was also more redshifted than the rest of the emission, suggesting it was orbiting around something heavy in CR7. All this data suggests that this object could have a growing supermassive black hole but as yet no stars have formed.

Finding more of these objects will be key—and that's no mean feat. The problem is that as objects in the universe get more distant, the signatures we use to trace supermassive black hole activity and star formation get redshifted so much that we can no longer see them with visible light. In other words, the Hubble Space Telescope can only see so far before there's nothing left for it to see. It's a good job that NASA and the European Space Agency (ESA) have a plan to launch another mission by 2021, called the James Webb Space Telescope, which will look at the sky in infrared light. This telescope will allow us to see the light from more distant objects as it gets redshifted from visible wavelengths down into the infrared. The stakes, and the expectations, for James Webb are high enough as it is, without the added excitement of finally solving the astrophysics equivalent of the age-old question: what came first—the chicken or the egg?

We Don't Know More Than We Do Know

There's a popular way of thinking about what we know and what we don't know. First, there are the *known knowns*—things we know that we know: that the Earth is a sphere; that there are planets around other stars; that the universe is expanding. Then there are things that we know we don't know—the *known unknowns*: we don't know what dark matter is made of; what form matter takes in black holes; or whether direct collapse black holes are possible. Then there are the *unknown unknowns*—things we don't know that we don't know. Hindsight gives us a few examples of these, such as the discovery of radioactivity after Marie Curie's experiments with uranium, or Benjamin Franklin's discovery of electricity. Thankfully, we can live in ignorance of the things we currently don't know we don't know while looking forward to the excitement of future discoveries that will, once again, change the human landscape on Earth.

My favorite by far is the fourth category—the *unknown knowns*: things that we don't know that we know. This category absolutely fascinates me. What are the things that we already have the knowledge or the tools to understand, but don't yet? For example, before we knew that galaxies were separate islands of stars outside our Milky Way, a meticulous chap called Messier, in 1771, classified all the fuzzy things in the sky that weren't stars. The fifty-eighth object in his extremely detailed list was actually a galaxy that was 68 million light years away. This was the most distant observed object in the universe at the time; he just didn't know it.

Another beautiful example of unknown knowns is Steve. Steve is a lesson for all of us in the modern world where the entirety of human knowledge is only a simple click away. Through 2015 and 2016, a group of amateur astronomers and keen astrophotographers called the Alberta Aurora Chasers had been capturing images of an interesting variety of Northern Lights in the sky that they hadn't seen before. It was a long, whitish-purplish streak running from east to west across the sky. It clearly wasn't a normal aurora of shimmering green and pink ribbons; such aurorae are caused by high-energy electrons burped off by the Sun in a solar wind that gets funneled by the Earth's magnetic field down to the poles where they exchange energy with elements in the atmosphere, causing them to glow. The Alberta Aurora Chasers decided this streak must be a different kind of aurora, one caused by high-energy protons rather than electrons in the solar wind hitting the Earth. So, they kept referring to these streaks across the sky as proton arcs.

The Northern Lights.

Later, they showed some of these images to a professional astronomer at a conference on the aurora. This expert, Eric Donovan, had spent twenty years studying aurorae and had never seen anything like the streaks in the images the group showed him. He immediately said it couldn't be a proton arc because aurorae caused by protons don't give off visible light—so we can't see them with our eyes. The Alberta Aurora Chasers, therefore, dubbed these strange streaks Steve, inspired by the children's animated film *Over the Hedge*, whose characters call things they're scared of, or don't understand, Steve.

As experts investigated Steve further they discovered that it wasn't a rare phenomenon at all and can, in fact, be seen farther south than typical Northern Lights. The problem was that the experts studying this were doing so with only two all-sky cameras across Canada to spot when aurorae appeared. By chance, they'd never had the opportunity to study a Steve with satellite data. We still don't fully understand what a Steve is and what causes it to appear in the sky, but research is ongoing in the hope that citizen scientists around the world will help by reporting when they see a Steve in the sky.[1]

I love this story, firstly because it showcases how anybody in the world can make a scientific discovery, but also because it teaches us that we should never assume that we

---

[1]   Check out NASA's aurora reporting project at aurorasaurus.org.

already know everything there is to know. The citizen scientists of the Alberta Aurora Chasers, after spotting Steve in the sky, assumed it was a known phenomenon—perhaps you, reader, have seen it and thought the same. It's a trap that we can all fall into—assuming that, in the twenty-first century, everything is already known and presumably documented somewhere on the internet. There is always room for us to learn and learning definitely doesn't stop outside the classroom.

Another classic story about unknown knowns—and another example of a scientific discovery originating from citizen scientists rather than "experts"—is that of Hanny's Voorwerp. Back in 2007, astronomers launched a website called Galaxy Zoo, which called on the public to help classify the shapes of over a million images of galaxies. It was a complete success and more than three hundred thousand people[2] worldwide got involved to help with this cutting-edge science. Prior to this project, these images had just been sitting on a computer hard drive—primarily because there simply aren't enough experts in the world to go through that much data themselves, so anyone logging on to the site stood a chance of being the first-ever human to see the image of that galaxy (this is

---

2 Galaxyzoo.org is still going and needs people to continue classifying for it. As astronomers, we never stop taking images of the sky so we will always need help reviewing them.

the danger of Big Data—it's a buzzword across most areas of science these days, but the result can be that needles go undiscovered in haystacks).

One of the volunteers on Galaxy Zoo was a Dutch schoolteacher called Hanny van Arkel. While classifying the shapes of galaxies, Hanny came across one image that had a fuzzy blue smudge underneath the galaxy. She was curious enough about it that she flagged it on the website's forum and asked what it was. The Galaxy Zoo team's experts were stumped. They'd never seen anything like it before. They didn't know if it was a real object, or whether it was something that had gone wrong while snapping the image. If it was real, they didn't know if it was in the foreground in our own Milky Way, or at the same distance as the galaxy, or whether it was in the background behind the galaxy in the image.

The first job was to confirm it was real—it turned out it was—and then to estimate the distance of light from both the galaxy and the blue smudge using the redshift of the spectrum. The redshifts of the two objects were the same and so the team then knew both galaxy and smudge were at the same distance as each other. A Hubble Space Telescope image of the smudge revealed it to be a cloud of gas with a complex structure that was very rich in oxygen, hence the blue glow in the original image. A complex cloud of glowing gas is a rather strange thing to find outside a galaxy in outer space—especially one that is glowing because of oxygen, as it takes a lot of energy to make oxygen glow.

The science team eventually deduced that the galaxy in the image had a much smaller companion galaxy orbiting around it. That companion galaxy had swept past the larger galaxy and had a gravitational interaction with the larger one. The forces involved in that interaction stripped gas off the companion galaxy in a long tidal tail. The larger galaxy was affected by those gravitational forces, too—it disrupted the very center of the galaxy and caused material to fall toward the supermassive black hole. That black hole was too greedy and had to throw some of the material out of the larger galaxy in huge jets traveling almost at the speed of light. These jets impacted the gas stripped off the smaller galaxy and caused the oxygen in it to glow. By this point, though, the supermassive black hole at the center of the galaxy was no longer actively growing, so we couldn't see that it was there. We call this a *quasar light ionization echo*—an echo because it shows us that the supermassive black hole was once active, but no more.

The wonderful thing is that, as soon as volunteers on Galaxy Zoo knew they were looking for fuzzy blue smudges in images, they found about forty more in the original set of a million images. That meant the experts had a full sample they could go away and study. None of this would have been discovered if it wasn't for the curiosity of one person: Hanny. After she posted the object on the forum asking what it was, other users started referring to it as "Hanny's Voorwerp." So, these quasar light ionization echoes are now commonly known as *Voorwerpjes* in astronomical journal articles—a fantastic word in its own right, fitting easily into the astronomical lexicon of quasars and quarks. The direct translation of *Voorwerp*, however, is "thingy," which is possibly my most favorite "thing" about this story.

One thing Hanny didn't know when she first posed the question about the strange Voorwerp she had seen is the result that question would lead to. It's something we call *blue sky* research—research that is not driven by a need for something, like a cure for a disease or a solution to a problem, but experimentation for experimentation's sake.

It's often how we justify astronomy research when confronted with the collective taxpayer cry: What have astronomers ever done for us?! How does asking questions about our place in the universe benefit humanity? To my mind, satiating curiosity is as good an answer as any, but often there are other, unforeseen advancements to collective human knowledge or technology that prove invaluable. The imaging techniques and cameras developed for us to see fainter and farther in the universe are now applied in medical imaging scanners to diagnose a whole host of ailments. The digital detectors invented to replace photographic plates, now employed in mobile devices we all carry in our pockets, were first developed to give astronomers more precise ways to measure how bright an object appeared. Those same pocket-dwelling devices also benefit from the methods used to boost and improve Wi-Fi signals developed by radio astronomers who needed to improve their data-transfer capabilities. These are technological advancements that most of us would comfortably state we would now struggle to live without.

So, although the topics covered throughout this book may have seemed otherworldly and beyond the scope of the everyday, the push to find the answers to the questions posed has enriched our day-to-day lives beyond reckoning. It would be remiss of us to assume that everything is now known and that we should halt our pursuit to understand space. There are still far more things that we don't know than we do know. And this means that we can look forward to the privilege of gaining more knowledge and understanding about the universe we call home.

As children, we have an innate sense of inquisitiveness that, somehow, seems to get lost as we grow into adults. Perhaps if we all took some time out from the hectic pace of twenty-first-century life to merely gaze up at the night sky and contemplate all the things we still don't know, we could once again feed that natural childlike curiosity in us all. Because it is curiosity that is the true driver of all science. Without it, we'll never be able to understand the true majesty, complexity, and mystery of the universe in all of its glory.

Acknowledgments

Two years ago you would have heard me confidently state that I would never, in a billion years, write a book, because how in the universe did people have enough patience to sit down and write an *entire* book?

In part, my barrier was that I was told by my high school English teacher that I wasn't a good writer—or, at least, that I write the way I talk. Turns out writing the way you talk might actually be a good thing for a popular science book. What I didn't realize is that when you write a book you get editors who help make you sound brilliant. So, to Emily, Anne, and Jennifer at Orion—thank you for turning my science word vomit into something elegant and concise. And for dealing with my love affair with *just*, a word that previously peppered this manuscript. To Shaida, Windy, Betsy, Dan, Mary, and all those at Penguin Random House for taking on the mammoth task of getting this published in North America. I'm always in awe of those who can create something out of nothing, so a big thank you to Justin for the beautiful artwork throughout this book.

I have collected all of the information in this book over many years of education from some truly wonderful people. Firstly, thank you to my thesis supervisor, Chris, and friendly neighborhood postdoc (now hotshot lecturer!) Brooke, who taught me that science is all about asking the right questions and that observations should never be left buried in desks. To my masters supervisor, Russell, who taught me that making mistakes is how we learn. To all my teachers at school, especially Mrs. Kyle, Mrs. McCann, and Mr. West, who taught me the foundations of math and physics. And, while I'm here, thank you to my first grade teacher, Mrs. Dean, for taking an inquisitive (and, let's face it, annoying) child into your class and nurturing her curiosity rather than smothering it. You should all know that everything I do, I do standing on your shoulders.

Thank you to my mum, dad, and sister for teaching me to be different, to look before I leap, and to dare to dream.

And finally, to Sam, for teaching me to laugh, *always*.

About the Author

Dr. Becky Smethurst is an astrophysicist and research fellow at the University of Oxford. Her current research is focused on the question of how galaxies and black holes evolve together. Her weekly YouTube videos explain unsolved cosmological mysteries, weird objects found in space, and general space news. Dr. Smethurst also presents physics videos for the YouTube channel Sixty Symbols and astronomy videos for Deep Sky Videos. She was shortlisted for an Early Career Physics Communicator Award by the Institute of Physics and was named Audience Winner in the UK national finals of the FameLab competition.

Index

Published by arrangement with The Orion Publishing Group Ltd. First published
in the United Kingdom in 2019 in slightly different form as *Space: Ten Things You
Should Know*.

Library of Congress Cataloging-in-Publication Data is on file with the publisher.

Hardcover ISBN: 978-1-9848-5869-6
eBook ISBN: 978-1-9848-5870-2

Printed in the United States of America

Design by Betsy Stromberg

2nd Printing

First US Edition

All numbers quoted in this work are accurate as of July 2019.